HOW TO SAVE ON YOUR ELECTRICITY BILL WITH SOLAR ENERGY

How to Save on Your Electricity Bill with Solar Energy

Walter the Educator

Silent King Books
A WhichHead Entertainment Imprint

Copyright © 2024 by Walter the Educator

All rights reserved. No part of this book may be reproduced in any manner whatsoever without written permission except in the case of brief quotations embodied in critical articles and reviews.

First Printing, 2024

Disclaimer

The author and publisher offer this information without warranties expressed or implied. No matter the grounds, neither the author nor the publisher will be accountable for any losses, injuries, or other damages caused by the reader's use of this book. Your use of this book acknowledges an understanding and acceptance of this disclaimer.

How to Save on Your Electricity Bill with Solar Energy is a little problem solver book by Walter the Educator that belongs to the Little Problem Solver Books Series. Collect them all and more books at WaltertheEducator.com

LITTLE PROBLEM
SOLVER BOOKS

INTRO

In recent years, the rise in global energy demands, environmental concerns, and technological advancements have driven many households and businesses to explore alternative energy sources. Solar energy, in particular, has emerged as one of the most popular and practical solutions to reduce dependence on traditional fossil fuels. Beyond the environmental benefits, the most immediate and tangible advantage of solar energy is the potential to save on electricity bills. This little book will delve deeply into how solar energy systems work, the types available, their financial and environmental benefits, and how individuals can maximize savings on their electricity bills by switching to solar power.

How to Save on Your Electricity Bill with Solar Energy

Understanding Solar Energy and how it Works

Solar energy harnesses the power of the sun and converts it into electricity through the use of photovoltaic (PV) cells. These cells are made from semiconductor materials, often silicon, which generate an electric current when exposed to sunlight. This process is known as the photovoltaic effect.

How to Save on Your Electricity Bill with Solar Energy

The primary components of a solar power system include:

1. **Solar Panels (Photovoltaic Modules):** These panels capture sunlight and generate direct current (DC) electricity.

How to Save on Your Electricity Bill with Solar Energy

2. **Inverter:** Converts the DC electricity into alternating current (AC), the type of electricity used by most household appliances.

How to Save on Your Electricity Bill with Solar Energy

3. **Mounting Equipment:** Holds the solar panels in place, typically installed on rooftops or in open fields.

How to Save on Your Electricity Bill with Solar Energy

4. **Battery Storage (Optional):** Allows excess electricity to be stored for later use, particularly useful for nighttime or cloudy days.

How to Save on Your Electricity Bill with Solar Energy

5. **Electrical Panel:** Distributes the solar-generated electricity to the household.

How to Save on Your Electricity Bill with Solar Energy

6. **Metering System:** Measures the electricity produced and tracks how much of it is used versus how much is sent back to the grid.

How to Save on Your Electricity Bill with Solar Energy

When solar panels generate more electricity than is needed by the home, the surplus electricity can often be sent back to the local utility grid. This results in credits on the household's electric bill through a system known as net metering, reducing costs further.

How to Save on Your Electricity Bill with Solar Energy

Types of Solar Power Systems

Before diving into how solar energy can reduce electricity costs, it's important to understand the different types of solar systems available. The choice of system can impact the degree of savings a household or business experiences.

How to Save on Your Electricity Bill with Solar Energy

1. **Grid-Tied Solar System:** A grid-tied system is connected to the local utility grid. This means that the household uses solar power during the day and can still draw electricity from the grid at night or during cloudy days.

How to Save on Your Electricity Bill with Solar Energy

The benefit of this system is that it allows for net metering, where excess electricity produced by the solar panels is fed back into the grid in exchange for credits. These credits can offset electricity drawn from the grid when the solar system isn't generating power. This type of system tends to be the most cost-effective because it doesn't require battery storage, which can be expensive.

How to Save on Your Electricity Bill with Solar Energy

2. **Off-Grid Solar System:** An off-grid system is completely independent of the local utility grid. Instead, the system relies on battery storage to ensure electricity is available during nighttime or periods of low sunlight. While this type of system offers complete energy independence, it is usually more expensive because it requires additional components such as batteries and charge controllers. However, off-grid systems are essential for remote areas where utility connections are unavailable.

How to Save on Your Electricity Bill with Solar Energy

3. **Hybrid Solar System:** A hybrid system combines the advantages of both grid-tied and off-grid systems. It remains connected to the utility grid while also incorporating battery storage. In times of high-energy consumption, the household can draw electricity from the grid, while excess solar energy can be stored in batteries for future use.

How to Save on Your Electricity Bill with Solar Energy

Financial Benefits of Solar Energy

1. Reducing Monthly Electricity Bills

The most immediate and noticeable financial benefit of solar energy is the reduction in monthly electricity bills. Traditional utility bills fluctuate based on energy consumption and the current market price for electricity, but once a solar system is installed, a significant portion of a household's energy needs can be met directly by the sun. In sunny regions, a well-designed solar system can generate enough power to cover 100% of a household's energy usage, resulting in zero utility bills or even negative bills due to net metering credits.

How to Save on Your Electricity Bill with Solar Energy

For example, a household with a monthly electricity bill of $150 might install a solar system that reduces their bill to $30 or less, depending on system size and efficiency.

How to Save on Your Electricity Bill with Solar Energy

Over the course of a year, this amounts to substantial savings. Over a 20-year period, these savings could easily exceed the initial investment in the solar system.

How to Save on Your Electricity Bill with Solar Energy

2. Protection from Rising Energy Prices

One of the main advantages of solar energy is that it protects homeowners from rising energy prices. The cost of electricity has been increasing steadily due to growing demand, inflation, and the rising costs of fossil fuel extraction. Solar energy allows individuals to lock in a portion or all of their energy costs, providing a hedge against future price increases. Once the solar system is installed, the cost of generating electricity remains consistent, as sunlight is free.

How to Save on Your Electricity Bill with Solar Energy

3. Tax Incentives and Rebates

Governments around the world recognize the environmental and economic benefits of solar energy, which is why they often offer incentives and rebates to encourage the adoption of solar technology.

How to Save on Your Electricity Bill with Solar Energy

In the United States, for instance, the federal government offers the **Investment Tax Credit (ITC)**, which allows homeowners to deduct a percentage of their solar installation costs from their federal taxes. In 2021, the ITC allowed for a 26% deduction, and though the percentage is set to decline in the future, it remains a substantial incentive.

How to Save on Your Electricity Bill with Solar Energy

In addition to federal tax credits, many states and local governments offer additional incentives such as rebates, performance-based incentives (PBIs), and renewable energy certificates (RECs), all of which can significantly reduce the upfront costs of installing solar panels. Combined, these incentives can lower the total cost of a solar system by 30-50%.

How to Save on Your Electricity Bill with Solar Energy

4. Increase in Property Value

Installing solar panels can also increase the value of a home. Studies have shown that homes equipped with solar energy systems sell for more than similar homes without solar systems. This is because potential buyers recognize the long-term cost savings associated with solar energy and are willing to pay a premium for a property that comes with reduced or eliminated electricity bills.

How to Save on Your Electricity Bill with Solar Energy

In regions where solar power is particularly popular, such as California, homes with solar panels have been found to sell 20% faster than non-solar homes.

How to Save on Your Electricity Bill with Solar Energy

Environmental Benefits of Solar Energy

While financial savings are a major reason for adopting solar energy, the environmental benefits should not be overlooked. Solar energy is a clean, renewable resource that significantly reduces a household's carbon footprint. This reduction in greenhouse gas emissions can contribute to the global effort to combat climate change.

How to Save on Your Electricity Bill with Solar Energy

1. **Reduction in Fossil Fuel Dependency:** Traditional electricity generation relies heavily on fossil fuels such as coal, natural gas, and oil. These energy sources are finite, and their extraction and use result in harmful environmental impacts, including air pollution, water contamination, and habitat destruction.

How to Save on Your Electricity Bill with Solar Energy

By switching to solar energy, households can reduce their dependence on these fossil fuels, leading to a cleaner, more sustainable energy future.

How to Save on Your Electricity Bill with Solar Energy

2. **Lower Greenhouse Gas Emissions:** Solar power systems do not produce carbon dioxide (CO_2) or other harmful greenhouse gases while generating electricity. Over the lifetime of a solar panel system, the reduction in CO_2 emissions can be substantial. According to estimates, the average residential solar system can offset 3 to 4 tons of CO_2 per year, which is equivalent to planting over 100 trees annually.

How to Save on Your Electricity Bill with Solar Energy

3. **Reduced Water Usage:** Many traditional power plants, especially those powered by fossil fuels and nuclear energy, require significant amounts of water for cooling. This can strain local water resources, particularly in arid regions. Solar energy systems, on the other hand, require minimal water for operation, helping to conserve this valuable resource.

How to Save on Your Electricity Bill with Solar Energy

4. **Promotion of Energy Independence:** By adopting solar energy, countries can reduce their reliance on imported energy, fostering energy independence. This reduces the vulnerability to international energy market fluctuations and geopolitical tensions that can affect the availability and price of fossil fuels.

How to Save on Your Electricity Bill with Solar Energy

How to Maximize Savings on Your Electricity Bill with Solar Energy

While switching to solar energy offers considerable savings, there are several strategies that homeowners can employ to maximize the financial benefits of their solar investment.

How to Save on Your Electricity Bill with Solar Energy

1. Optimize Energy Efficiency before Installing Solar Panels

Before investing in a solar system, it's wise to first reduce overall energy consumption by making energy-efficient upgrades. This can include installing energy-efficient appliances, using LED lighting, improving insulation, and sealing air leaks. By lowering your household's energy demand, you can reduce the size of the solar system needed, thus lowering upfront costs.

How to Save on Your Electricity Bill with Solar Energy

2. Size Your Solar System Correctly

The size of your solar system should be tailored to your specific energy needs. Oversizing the system can result in unnecessary costs, while undersizing it may not provide sufficient savings. A solar installer can assess your household's energy consumption patterns and design a system that meets your needs while optimizing savings.

How to Save on Your Electricity Bill with Solar Energy

3. Take Advantage of Time-of-Use Rates

Some utility companies offer time-of-use (TOU) rates, where the cost of electricity varies based on the time of day. Solar systems can help you avoid high electricity rates during peak hours by allowing you to use solar energy during these times or by storing excess energy in batteries for later use.

How to Save on Your Electricity Bill with Solar Energy

4. Monitor and Maintain Your Solar System

To ensure your solar system operates at peak efficiency, regular monitoring and maintenance are essential. Dust, debris, and shading from nearby trees can reduce the amount of electricity your system generates. Many solar inverters come with monitoring software that allows homeowners to track system performance in real-time and detect any issues early on.

How to Save on Your Electricity Bill with Solar Energy

5. Consider Solar Battery Storage

While not necessary for grid-tied systems, adding battery storage to your solar system can provide additional savings by allowing you to store excess solar energy and use it during times when electricity rates are high. Batteries can also provide backup power during grid outages, offering peace of mind and additional savings.

How to Save on Your Electricity Bill with Solar Energy

6. Leverage Government Incentives and Net Metering

To maximize financial benefits, homeowners should take full advantage of all available government incentives, including tax credits and rebates.

How to Save on Your Electricity Bill with Solar Energy

Additionally, enrolling in a net metering program can provide further savings by allowing homeowners to earn credits for any excess electricity they generate.

How to Save on Your Electricity Bill with Solar Energy

OUTRO

Solar energy offers an effective and increasingly affordable solution to reduce electricity bills and contribute to a more sustainable future. By understanding how solar systems work, selecting the right type of system, and taking advantage of government incentives and net metering, homeowners can achieve significant savings on their electricity bills. Moreover, the environmental benefits of solar energy, including the reduction of greenhouse gas emissions and reliance on fossil fuels, make it a compelling choice for individuals looking to make both a financial and ecological impact. With proper planning, maintenance, and efficient energy use, solar power can be a game-changer for reducing household energy costs and promoting a greener planet.

ABOUT THE CREATOR

Walter the Educator is one of the pseudonyms for Walter Anderson. Formally educated in Chemistry, Business, and Education, he is an educator, an author, a diverse entrepreneur, and he is the son of a disabled war veteran. "Walter the Educator" shares his time between educating and creating. He holds interests and owns several creative projects that entertain, enlighten, enhance, and educate, hoping to inspire and motivate you. Follow, find new works, and stay up to date with Walter the Educator™ at WaltertheEducator.com

Milton Keynes UK
Ingram Content Group UK Ltd.
UKHW021938281024
450365UK00018B/1154